I0563964

a gift to

_____

from

_____

# THE mind OF A CAT

writings from Liz Abeler Blaylock

KPT|PUBLISHING

## ABOUT THE AUTHOR

Liz Abeler Blaylock lives in a messy house in Minnesota with her husband, daughter, grandson, parents (downstairs in the "mother-in-law" apartment—and yes, their part of the house is not messy,) two goldfish, and a beautiful new rescue cat named Cali.

By day, Liz works with middle school special-needs students. By night, she makes dinner, does laundry, moving piles of stuff from one place to another. Her other books are *I am the Cat* and *I am the Dog*.

> *"I marvel at the intricate, beautiful, and fearful working of creation, and revel in the grace and mercy of the Creator."*

Enjoy my memories!

I like
little people.
They can see
*exactly*
what I see.

Hey!
We're like twins—
but you smell
like fish.

I await further
instruction…
that *I want*
to hear.

Smile?

*I am* smiling.

How does
it do that?

This one is
*juuust* right.

What do you mean,
"Princess
and the Pea?"

I'm litterbox trained.

Decorations are ready for the party...

What?! This *isn't* the "party" room?

Can I have
a treat?

Can't imagine where
all my toys went.

I'm thinking
about mischief
unless you have
other plans.

My favorite
place ever...

*at least for today.*

The portal
to contentment.

We cats have been
a treasured mystery
for millenia.

Another
treasured mystery:
When is dinner?

If we had thumbs,
we'd give you
eight thumbs up!

This *better* be
important.
I was dreaming
about string.

EVERY day
is a
*CAT*-urday!

No two cats
are exactly alike.
In fact,
no *one* cat
is exactly the same
from day to day.

Without me,
you'd just be
a crazy lady.

I'm not big on words
*unless* it's
"Kitty" or "Tuna."

Let me help
with this.
I'm pretty good
with a mouse.

I *am* helping…

letting you know
these are clean
and soft and warm.

Yes,
I'm one in
600 *million*.

Have a nice bath,
you said.

It will be relaxing,
you said.

I was *expecting*
a lot more licking.

I guess I got
a little
carried away.

**The Mind of a Cat**

© 2018 KPT Publishing, LLC
Written by Liz Abeler Blaylock

Published by KPT Publishing
Minneapolis, Minnesota 55406
www.KPTPublishing.com

ISBN 978-1-944833-49-7

*Designed by AbelerDesign.com*

First printing November 2018

10 9 8 7 6 5 4 3 2 1

Printed in the United States of America